General Construction Engineering

Priti Dave

Published by Priti Dave, 2024.

While every precaution has been taken in the preparation of this book, the publisher assumes no responsibility for errors or omissions, or for damages resulting from the use of the information contained herein.

GENERAL CONSTRUCTION ENGINEERING

First edition. July 3, 2024.

ISBN: 979-8227205858

Written by Priti Dave.

ISBN : 9781548524449

First Published : July 2017

GENERAL CONSTRUCTON ENGINEERING

Self help (Reference guide)

BY

<u>Priti Dave</u>

Introduction

MRS. PRITI U. DAVE is Civil Engineer from Shri Bhaghubhai Mafatlal Polytechnic Mumbai (India). She believes 'sharing will enrich everyone with more knowledge ' The Idea behind writing the book ' General Construction Engineering' is to integrate all the major activities required to be checked at the time of project execution. In her long work tenure she acquired quality experience and knowledge. Efforts have been made in this book to provide information useful to beginners and professionals working on construction sites.

This book contains two sections.

Section – I Describes checklists required for various construction activities.

Checklists at various stages of project execution is useful guide for all technical professionals at site to perform the work effectively.

In addition, machineries and equipment require and registers to be maintained at site. Elevator work, fire-fighting work, construction of bituminous roads such specialized works are required to be done systematically. Checklists for all these works are very useful at the time of construction.

Standard values/measurements/dimensions are provided for instant reference to technical professionals working at the site.

Section – II is detailed with useful formulae and data required at various stages of construction work.

In Construction the calculations required to estimate quantities, to obtain test results, to work out strength of concrete, reinforcement requirements for structural members, labour strength etc. For different construction works. Appendix -I to XII are detailed with data useful to technical professionals in construction industry.

Contents
Section – I Checklists

SR. NO. SUBJECT PAGE no.

1. Lists of documents required at the time of

buying property/land 2

1. Approvals required from government

authorities 4

1. Lists of drawings and information to be collected

From Architect 5

1. Lists of drawings and information to be collected

From R.C.C. consultant 9

1. List of drawings and information to be collected from

Plumbing consultants 11

1. Lists of drawings and information to be collected from

Electrical consultant 12

1. Major heads to be considered while preparing work order

Or agreement for various contractors
R.C.C. work 14

Masonry work 16
Plaster work 16
Flooring work 17
Waterproofing work 18
Plumbing work 19
Electrical work 20
Painting work Road work

1. Lists of machineries and equipment 23

2. Checklists for construction activities at various stages 27

3. Check lists for Plumbing work 59

1. Checklists for Drainage work 60

2. Checklists for Aluminium window fixing 61

3. Checklists for Electrical works 62

4. Checklists for Elevator works 63

5. Checklists for Fire -fighting work 65

6. Check lists for Bituminous Road work 66

7. Lists of registers to be maintained at site 67

8. Standard values/measurements/dimensions 68

Section - II
<u>Calculations in construction</u>

1 Concrete quantity for footing   73

2 Concrete quantity for column   74

3 Concrete quantity for Beam   74

4 Concrete quantity for slab   74

5 Target mean strength of concrete   75

6 Determination of concrete density   76

7 Determination of cement content in concrete   77

8 Water content in concrete mix   78

9 Aggregate content in concrete mix   78

10 Individual aggregate content in concrete mix   79

11 Weigh batching for conventional concreting   80

12 Volumetric batching for conventional concreting 81

13 Bulking of sand   82

14 Silt content of sand   83

15 Water absorption test on metal 84

16 Water absorption test on bricks & blocks 85

17 Water absorption test on mosaic tiles 86

18 Steel quantity (Reinforcements) 87

19 Slenderness limits for beams 88

20 Cover to Reinforcements 89

21 Reinforcement requirements for Beam 90

   Reinforcement requirements for Slab 90

   Reinforcement requirements for column 90

22 Calculation of labour strength 94

23 Appendix-I Weight of Building materials 95

24 Appendix – II Geometric formulae 97

25 Appendix- III Weight of Reinforcement Bars 99

26 Appendix – IV Weight of structural steel 100

27 Appendix – V weight of M.S. angle 101

28 Appendix – VI R.C.C. steel in KG/cubic feet of concrete 102

29 Appendix – VII Consumption factors

   For cement 103

   For Sand and aggregates 107

   For Miscellaneous items 108

30 Appendix- VIII Concrete mix volumetric proportion 110

31 Appendix – IX General Conversion table 111

32 Appendix – X Units of measurements 113

33 Appendix- XI Labour task work 115

34 Appendix – XII Labour requirement for different works 116

BUYING A PROPERTY/ LAND/PLOT

List of Documents

1 NOC from Electricity Dept.

2 NOC from Pollution Dept.

3 NOC from Water Works dept.

4 NOC from Port Authority

5 Sale Deed,

6 Title Deed,

7 Mother Deed,

8 Conveyance Deed

9 Joint Development Agreement

10 Power of Attorney

11 Sanctioned Building plan

12 RTC Extracts

13 Katha Certificates and Extracts

14 Mutation Registration Extracts

15 Sale & Construction agreement between developer/Builder & 1st owner

16 Copy of Possession letter from Builder / Developer

Any Loans on Property verification

———◦———

17 COMPLETION CERTIFICATE

18 Occupancy Certificate

19 Deed of Declaration

20 Latest Electricity Bill

21 Sale Agreement with the Seller

22 All Paid Tax receipts

23 EC up to date for last 13 years

24 Demand letter from Vendor before disbursement

25 NOC from Society / Building

26 No Due Certificate

27 Approved Plan

28 Layout approval plan

29 Auction Sale confirmation letter from local authority

30 Release Deed if any

31 Power of attorney from landlord in the name of builder through which he is empowered to develop the land.

<u>BUILDING APPROVALS FROM GOVERNMENT AUTHORITIES</u>

1 N.A. (Non- Agriculture) use certificate

2 Commencement certificate

3 Drainage approval

4 Road approval

5 water connection

6 Lift/Elevator (N.O.C)

7 Urban Land Ceiling (U.L.C) (N.O.C.)

8Occupancy certificate

4

ROLE OF ARCHITECTS

A. ARCHITECTS

1. Area calculations

1. Plinth area

2. Floor Area

3. Carpet area

4. Built up Area

5. Super Built up area

6. Parking area

2. List of Drawings

I Demarcation drawings

II Municipal drawings

III Layout drawing

IV Basement drawings

V working drawings for

1. site office

2. watchman cabin

3. cement go-down

4. Labour hutments

5. material store rooms

6. STEEL YARD

7. water storage tank

8. Water supply to different structures

9. Electric supply to different structure

10. Transformer,L.T.Room and Feeder pillar

location details

11. Internal Road connecting approach road

12. Compound wall details

13. Main Gate Details

14. U.G.tank water Tank/Pump room details

15. Septic tank Details

VI Building wise Floor Plan

1. Ground floor plan/stilt floor Plan

2. First Floor/ Typical Floor Plan

3 Terrace Floor Plan

4Lift Machine room details

5 Over Head water Tank details

6 Section through staircase drawing

7 Front & Rear side Elevation drawing

8 Doors/windows/rolling shutter detailed drawings

10 Kitchen Platform details

12 Toilet details

13 Painting details

3. N.O.C. Clearance from Architect

1. Line-out for each structures at the site

2. Line-out for each building upto plinth

3. Concreting for each slab

4. Sample work of all finishing items

5. Revision drawings approval from municipal authorities

6. Perspective drawing of the project

7. colour scheme of the project

8. Tender documents

9. Model of project

10. Storm water drawing

11. Drainage drawing

12. Water supply lines

13. Plumbing fittings

14. Fire- fighting system

15. Swimming pool details/Filtration plant details
16. Electrical layout plan in the premises
17. Road lights position
18. cable laying below road
19. Cable sizes & Layout
20. Earthing position of each building
21. Busbar specifications
22. Conduit details for TV/Telephone wiring
23. Satellite dish antenna for Project

<u>ROLE OF STRUCTURAL CONSULTANTS</u>

B. R.C.C. Structural Consultant

I List of Drawings

1. Bearing capacity of the soil from trial pits
2. Basement Raft slab detailed drawing
3. Retaining walls/side pardies detailed drawing
4. Footings drawings
5. Columns schedule
6. Floating columns schedule
7. Plinth beams schedule & profile drawing
8. tie beams Schedule & profile drawing
9. Parking Slab schedule & detailed drawing
10. Typical floor slab Schedule & detailed drawing
11. Mezzanine slab / ground floor slab detailed drawing
12. Terrace level slab detailed drawing
13. Teraace level & Balcony Pardhi detailed drawing
14. Lofts/Lintels & Chajjas detailed drawing
15. Staircase detailed drawing
16. Overhead water tank & Machine room detailed drawing
17. Lift shaft bottom slab detailed drawing
18. Machine foundation detailed drawing

---

19. U.G. TANK/SEPTIC tank detailed drawing
20. R.C.C. grill details
21. M.S. staircase & Ladders details
22. Expansion Joint details
23. Decorative RCC work details

II. N.O.C. & clearence from R.C.C. consultant

1. Laying foundation P.C.C.
2. Footing & Stub column concreting
3. Raft concreting

4. Plinth filling

5. concreting at plinth level

6. column concreting at every stage

7. slab concreting at every stage

8. Lofts / chajjas/pardies concreting

9. O.H. Tank/ U.G.Tank/Swimming pool concrete works

10. Submitting Stability certificate of the project to government authorities

ROLE OF OTHER CONSULTANTS

C. PLUMBING CONSULTANT

N.O.C. by the plumbing consultant

1. Plinth & Parking levels of the building
2. work of sewer line
3. Concealed G.I. line work
4. Open G.I. line work
5. Sample Bathroom Unit
6. Sample Toilet Unit
7. sample Kitchen plumbing work
8. Lead & other joints in piping work
9. Ground drainage network (Manholes& chambers)
10. septic tank & soak pit
11. terrace G.I. line work
12. Storm water systems for entire project
13. O.H. water Tank pipe connections
14. U.G. water tank pipe connections
15. Pump requirement as per standard quality

D. Electrical Consultant

1. Basic power requirements

2. L.T. substations location & numbers

3. Basic specifications for Transformers

4. Main panels

5. Drawings for

L.T.substation

H.T. (overhead& Underground) distribution lines

L.T. (overhead & underground) distribution lines

sports ground lighting

Garden lighting

Lighting for recreation facilities

Lighting for swimming pool

cable schedule

Earthing layout drawing

AGREEMENTS WITH CONTRACTORS

<u>**NOTE: HEADS BELOW LISTED**</u> are to be detailed as per specifications and concern with contractor , client , technical professional at project site.

A. Agreement For Civil works includes following heads

1. Name of contractor's firm & address
2. Scope of work
3. Project name &address
4. Estimated cost of the work
5. Quotation for the work from contractor
6. Technical specifications for the work
7. Bill of Quantities for the work
8. Schedule of Rate per item of the work

<u>R.C.C. Works</u>

Points to be considered while preparing agreement is as follows

1. Mode of measurement of all R.C.C .work

2. Progress report

3. Workmanship

4. Defect liability period

5. Quality of materials

6. Supervision of the work

7. Labour wages

8. Subcontracting the work

9. Mode of payment

10. Shuttering of R.C.C. work

11. Scaffolding for all R.C.C. work

12. Steel Reinforcement

13. Binding wire

14. Oiling shuttering planks,steel plates etc.

15. Repair of Honeycombing on concrete works

16. De-shuttering period for all R.C.C. work

17. Curing all R.C.C. work

18. Plinth filling & Masonry upto plinth level

19. Railing all around the building

20. HACKING ALL R.C.C. work

21. Cover Blocks for all R.C.C. work

22. Labour insurance

23. Safety helmets/shoes/gloves

24. Inspection request to site in charge

25 Labours insurance

<u>Masonry & Plaster works</u>

Points to be considered while preparing agreement is as follows

1 Scope of work

2 Schedule of Rate

3Line -out per floor

4 Sufficient wet Bricks/Block

5 Fixing door & window frames

6Fixing Aluminium frame

7 Mortar joint (half inch)

8Curing all Masonry & plaster work

9 Plastering wall/ceiling on wet surface

0 Sand screening & washing

11Inspection request to site in charge

12 Skirting work

13 Water spouts in balcony

15Scaffolding

16 chicken mesh fixing

17 Butt/Jad finishing for skirting/dado/kitchen platform etc.

18Mode of measurement for Masonry & Plaster work

19Gabhadi/touch up work

<u>Flooring/Tiling work</u>

Points to be considered while preparing agreement is as follows

1 Scope of work Schedule of rate

2Stacking of tiles as per shades

3Tools& Machineries

4Bedding Mortar

5Fixing of the tiles

6Wastage

7To Maintain required slope

8 Tile cutting

9Joint filling

10Curing

11Skirting

12Steps & Risers

13Polishing

14 Cement grouting of tiles

15 Inspection Request to Engineer -Incharge

16 Defects in work

17Acid washing

18 Fixing Kitchen otta/platform

19 Mode of measurement

20 Mode of payment

<u>WATER PROOFING WORK</u>

Points to be considered while preparing work order is as follows

1 Socketing and cement grouting

2 Base coat

3 water escape pipe

4 Brick bat coba

5 Final topping coat

6 Curing

7 Terrace/kitchen sink/O.H.water tank/U.G. water tank water proofing

8 Ten years Guarantee

9 Maintenance of the work

10 Mode of measurement

11 Mode of payment

PLUMBING WORK

Points to be considered while preparing work order is as follows

1 Scope of work

2 Manpower and tools

3 License for work

4 Cutting and threading cost

5 Lead joints

6 Coal tar/Hessian cloth

7 Line out

8 Finishing

9 Scaffolding

10 External Plumbing

11 Internal plumbing

12 Testing

13 Leakage rectification

14 Rates

15 Curing

16 Retention

17 Standard heights of fittings

18 Maintenance of work

19 Estimate for work

20 Mode of measurement

21 Mode of payments

<u>ELECTRICAL WORK</u>

Points to be considered while preparing work order is as follows

1 Work as per I.S. Electricity Board
2 Materials Brand/specifications
3 Scope of work
4 Bills
5 Retention
6 Testing of works
7 Electrical points
8 Boards and Boxes
9 Conduits
10 Supply cables
11 Meter cabinet wiring
12 Maintenance
13 Estimate of work
14 Mode of measurement
15 Mode of payment

PAINTING WORKS

Points to be considered while preparing work order are as follows

1 Scope of works

2 Material specification

3 Maintenance of work

4 Quantity of work per day by skilled labour

5 Rates

6 Mode of payment

7 Mode of measurement

ROAD WORKS

Points to be considered while preparing work order are as follows

1 Scope of work

2 Machineries

3 Drawings

4 work stages

5 Procurement of material

6 Consumption of Bitumen

7 Levels/slopes/cambers of road

8 Material specification

9 Rates

10 Mode of measurements

11 Mode of payment

12 Maintenance

<u>MACHINERIES,EQUIPMENTS</u> &
<u>INSTRUMENTS</u>

1 Steel cupboards

2 tables &chairs

3 Safety helmets/shoes/gloves

4 Rubber stamps & stamp pads

5 Company Name board

6 First Aid box

7 writing board

8Soft board

9 display board

10 Various shapes & sizes of nails

11 Kathya,rope, line dori

12 G.I.Pipes of various sizes

13 G.I. fittings(sockets,reducers,enlargers,elbows, bend tee,union plugs check nuts)

14 hammer,chisel,tacha,screw-drivers,spanners,steel tapes,right angle holdfast, aldrop

15 Ghamela, Phavada, bucket, iron bars, binding wires, ladders, drill machine.

16 Paint brushes of all sizes

17 Primer

---

18 RED OXIDE

19 Turpentine

20 Lime bags

21 Polish papers

22 Scrapper

23 Brooms

24 Coal tar

25Wires of different gauge

26 Bulb holderd
27 Adapters
28 Insulation tapes
29 Switches & sockets
30 Adhesives
31 Corrugated G.I. sheets
32 Hinges
33 Electric pups
34 Diesel pump
35 Vibrator
36 Weigh balance
37 Weigh batchers

38 Blocks making machine

39 Block pallets

40 Silt measuring flask

41 cube moulds

42 Slump cone apparatus

43 Sieve sets

44 Sand screening machine

45 Sand washing machine

46 Dumpy level

47 Theodolite

48 Plane table

49 Concrete buckets

50 Concrete pumps

51 Concrete mixers

52 Concrete needle vibrator

53 Trolleys

54 Pulleys

55 Dumpers

56 Props & Spans

57 Water Pumps

58 Electric motors
59 Road roller
60 Excavator
61 Bulldozer
62 Bitumen mixing plant
63 Tremix concrete machine
64 Rock cutting compressors
65 Lift & hoists

STAGES OF PROJECT EXECUTION

CHECK : LAYOUT : ALLOW Excavation

Refer Drawings : Site layout

Centre line Plan

Check the following

1 Overall dimension

2 All angles

3 Centre lines and grid distances

4 Diagonals

5 Distances from bench mark, plot boundries etc.

6 Whether pillars or side railings, with centre line marked/ Painted, have been provided.

7 whether levels have been transfered to site

8 Line out of excavation areas to ensure size, working space and orientation.

Check : Excavation : Allow P.C.C.

Refer drawings : Centreline plan

Foundation Plan

Structural details of foundations

Check the following

1 Pit Size

2 Depth of Excavation

3 Bearing Capacity of Soil

4 Shoring, If required

5 Pit is dry

6 Excavation bottom well compacted

7 Rubble soling

8 Centre line have been marked on side of excavated pit.

9 Level mark for top of P.C.C.

Check : Shuttering Allow : Concreting of Foundation

Refer drawings : centre line drawing

Structural details of foundations

Check the following

1 Level on the top of P.C.C.

2 centre lines marked on P.C.C. surface

3 Length , breadth , diagonals of shuttering

4 Depth of concreting

5 Reinforcement - diameter, numbers,length etc.

6 Cover to Reinforcement

7 Inserts, if any

8 Stability of formwork

9 Approved mix design

10 Materials - aggregates and sand are available as per desired quality

11 Approaches and proposed method of concreting

12 Support to column Reinforcement

13 Availability of vibrator

Check : Shuttering Allow : Stub Column

Refer Drawings : Centreline

Structural details of Stub column

Layout of columns

Plinth beam

Check the following

1 Centre line marked on foundation concrete

2 Orientation

3 Shuttering dimension length, breadth and depth

4 Reinforcement - diameter, number, stirrups et

5 Cover to Reinforcement

6 Support to column reinforcement

7 Stability of shuttering

8 Level upto bottom of plinth beam

Check : Shuttering : Allow : Plinth Beams

Refer Drawings : Plinth floor plan

Structural details of Plinth beams

Check the following

1 Centre lines

2 Level at top of plinth beam

3 Shuttering dimensions - length, breadth, and depth

4 P.C.C. on earth well compacted beneath beam, if shuttering is not to be provided

5 Stability of shuttering

6 Reinforcement - diameters, numbers, stirrups, bends, laps etc.

7 Cover to reinforcement

8 Inserts for toilet pipes etc.

9 Top of stub column is rough and free from all loose material and laitance

Check : Foundation/Plinth beam : Allow : Back filling in foundation & between
Plinth beams

Refer drawings : Floor plan giving Grade slab level

Check the following

1 No loose material within pit and between plinth beams

2 Level upto top of filling is marked

3 Bitumen painting, if specified, on footings and plinth beams

4 Anti-termite treatment, if specified

5 Material to be used for back - filling is well graded and boulders, timber and deteoriousmaterial are absent

6 225mm level marks painted on foundations and plinth beams. (Back-filling to be carried out in layers of 225mm, well- watered and compacted to 150mm)

Check : Rubble soling : Allow : Grade slab

Refer drawings : Plan/ Sections given levels

Check the following

1 Rubble soling well Placed and compacted

2 Grade of concrete to be used

3 Thickness of Grade slab

4 Level at top of slab , marked

5 If size of slab is above 20sqm. , it should be paid in panels, in chess board pattern.

check shuttering for panels

6 Anti-termite treatment , if specified.

7 Sunken portions for toilets, if any.

Check : Back-filling Allow: Rubble soling in Grade slab

Refer drawings : Plan/section giving finished levels

Check the following

1 Back -filling is well compacted and impervious to water. (Dry density test may be taken. If not specified press thumb - it should not depress surface)

2 Top level of soling , is painted on plinth beam sides

3 Rubble stone - these should be squarish in shape and between 175mm to 225mm in size

4 Availability of heavy compactor.

5 Availability of grit to be spread over soling and compacted.

6 All rubble to be pitched vertically with larger size of stone as base.

Check : Shuttering : Allow : Column starters

Refer drawing : Column Layout

Structural details of column

Check the following

1 Centre lines

2 External face of all outside columns to be in line

3 Orientation

4 Cover to reinforcement

5 Dimensions and diagonals

6 Surface cleaned of all loose material

7 Shuttering depth of 150mm

8 Richer mix of concrete to be used.

9 Concrete to be machine mixed.

Check and Allow : Rubble soiling

Refer drawings : Foundations

Floor plans

Check the following

1 The rubble should be of sound quality of even size (between 150mm and 225mm) and do not have narrow edges.

2 Sufficient headers stones are available. Each course to have headers at 1800mmC/C and are staggered.

3 the face area of each stone should be 0.35sft. minimum. For headersit should be 0.05sft.

Following to be ensured while Rubble masonry is in progress.

1 Mix as specified.

2 The broader side of stone is placed at the bottom

3 All gaps are filled with proper sized stone

4 Mortar must surround each single stone

5 Mortar joint does not to exceed 12mm

6 300mm high key stones are leftin the masonry at the end of days work.

7 300mm deep keys are left for joints in right angled walls

CHECK : SHUTTERING ;Allow: Column concreting

Refer drawings : Column Layout

Structural details of columns

Roof beams and slab above

Check the following

1 Starter cured for 24 hours

2 Honey combing and cracks in starter, absent

3 Reinforcement

Diameter and number of bars

Spacing of stirrups

Size and shapes of stirrups

Cover to reinforcement

Rust on steel

Laps, if any should be staggered

Lap length of dowels for columns of next floor

4 Formwork

Dimensions- length, breadth, diagonals and height

Column formwork to extend upto the exact bottom of beam above

Internal surfaces to be plumb

stability of formwork

Access for vibrator

Center lines

Column shuttering held in place by external bolts

5 Inserts if any

6 Method of placing concrete

7 Hand mixed concrete should not be allowed under any circumstances

Check : shuttering : Allow : Beams, Slab, chajjas , canopies

Refer drawings : Floor plan

Structural details

Check the following

1 Columns cured for three days before start of shuttering

2 7 days cube results of column concrete acceptable

3 Reinforcement

Diameters, numbers and spacing

Lap length

Bends

Chairs

Cover blocks

Binding wire well tied

For reduced column size above jogging of bars to be done within beam

4 Form-work

Dimensions - length, breadth, diagonals

Level of bottom and top of slab

Well oiled

All gaps sealed with galvanised iron strips

Wherever plywood/timber is used as filler it should be of full opening size of one piece

5 Stoppers, if required, provided correctly

6 Top of beam provided with spacer of shape

7 The inner side shuttering of beams, for sunken slabs, are supported by chairs and not timber supports

8 Supports

Props are vertical

Props are equally and suitably spaced

Props have lateral ties

Props have no joints

If timber props have been used, there should be of minimum dia of 75mm

The floor on which props rest are clean of all material

Adjustment of height is by timber wedges

Runners are of proper size

Side support to beams are rigid and stable

9 Electrical conduits, fixed as per drawings

10 Fan hooks

11 Inserts if any

12 Method of placing concrete

13 Lights for night concreting

14 Availability of Cement, Sand, aggregates and water

15 Availability of two poker vibrators and one screed vibrators

16 Mixer, hoist, trolly etc.

17 Slump cone and six cube moulds

18 Tarpaulins of adequate size during monsoon

Check and Allow : Block masonry

Refer drawings : Floor plans

Sections

Check the following

1 All concrete surface properly hacked (6mm deep and 30mm c/c)

2 Blocks are as per approved quality.

3 Location of doors and opening marked.

4 Door/Window frames are at site.

5 Number of layers determined

6 Number of blocks in a horizontal row, determined.

7 measuring boxes for cement and sand.

8 Blocks have been pre-wetted.

Following to be ensured while Block Masonry is in Progress

1 Mortar is mixed properly in the same proportion at all times.

2 All joints to be even and about 10mm thick.

3 Line , level and Plumb.

4 All vertical joints to be in plumb.

5 For 115mm thick walls P.C.C. band 150mm thick is cast at a height of every

1200 mm.

6 No gaps to be left in walls. Gaps, if any must be filled with concrete.

7 The top layer, beneath the beam, should have full thickness block and provision for

mortar 20mm thick. If not possible, then concrete the opening.

8 Mortar is placed between edge of block and column.

9 Not more than 1200mm height of masonry to be done in one day.

10 All joints are raked out properly.

11 Adequate curing is done

Check and Allow : Brick Masonry

Refer drawing : Floor plans

Sections

Check the following

1 All concrete surface properly hacked (6mm deep and 30mm c/c)

2 Bricks are as per approved quality.

3 Location of doors and opening marked.

4 Door/Window frames are at site.

5 Number of layers determined

6 Number of bricks in a horizontal row, determined.

7 Measuring boxes for cement and sand.

8 Bricks have been pre-wetted.

9 Concrete surfaces have been prewetted

Following to be ensured while Brick Masonry is in Progress

1 Mortar is mixed properly in the same proportion at all times.

2 All joints to be even and about 10mm thick.

3 Line , level and Plumb.

4 All vertical joints to be in plumb.

5 No Bricks, less than half in length is used.

6 No gaps to be left in walls. Gaps, if any must be filled with concrete.

7 The top layer, beneath the beam, should have full thickness brick and provision for

mortar 20mm thick. If not possible, then concrete the opening.

8 Mortar is placed between edge of brick and column.

9 Not more than 1200mm height of masonry to be done in one day.

10 All joints are raked out properly.

11 Curing Continues for seven days.

Check : Masonry Allow : Internal Plaster

Refer drawings : Floor plans

Schedule of finishes

Electrical layout

Plumbing layout

Check the following

1 Masonry Joints have been raked out.

2 Masonry cured for seven days.

3 All electrical conduits and boxes have been laid vertically, and recesses concreted.

4 All Plumbing pipes and fittings have been fixed,and recesses concreted

5 Expanded metal, 200mm wide, has been fixed over recessed conduits and pipes, joints of concrete to masonry, masonry to timber and timber to concrete.

6 Mix of mortar checked with specification. if not specified use 1:4 for walls and 1:3 for ceiling.

7 Walls have been prewetted, 24 hours in advance.

8 Thickness and level courses have been marked.

9 Door and window frames fixed

10 Aluminium straight edges 100mm x50mm, 2m long is available at site.

11 Motar mixer and measuring boxes are at site.

12 Cement grout has been spread on ceiling and cured for two days.

Following to be Ensured during plastering

1 Water is added to mortar by 5 litre tin and not by hose pipe.

2 Water cement ratio to be maintained at 0.7.

3 If hand mixing is allowed then a brick or block wall surrounding to be made to avoid loss of water while mixing.

4 No dry cement to be used.

5 Curing is done for 7 days.

Check : Masonry Allow : External plaster

Refer drawing : Elevations

Check the following

1 All concrete and masonry surfaces are in line and plumb.

2 The scaffolding self standing and rigid.

3 Window frames have been protected and wrapped with plastic sheets.

4 Thickness level marks have been made at a distance of 4mc/c. If plaster thickness is 20mm, the thickness marks to be made in two layers of 8mm and 12 mm.

5 Masonry has been prewetted 24 hours in advance.

6 200mm wide expanded metal fixed.

Following to be Ensured while plaster is in progress.

1 All mortar to be mixed in mixer. Material measured in boxes and water cement ratio to be maintained at 0.7

2 First coat to be mix 1:4 and second coat with mix 1:3 or as specified.

3 the first coat to be roughened with wire brush.

4 Drip moulds to be provided at all sills, chhajas, parapet walls etc.

5 Top of parapet wall to be sloping internally and internal parapet wall to be finished smooth up to height of 300 mm from slab.

6 All plumbing pipe openings etc. should be properly concrete and plastered.

Check and Allow : Mosaic floor tiles

Refer drawings : Floor plans

Floor finishes

Check the following

1 Centre of room has been marked.

2 Floor thoroughly cleaned, all loose material removed and uneven surfaces cleared.

3 Floor slopes checked and levels marked on floor and walls.

4 All tiles have perfect corners and are not damaged.

5 Floor level differences between rooms and toilets and room balconies to be 25mm.

6 Divider strips if any.

Following to be Ensured while tiling is in progress.

1 Mix is proper.

2 Joints are even, 1mm width.

3 Mortar is fully compacted beneath tiles

4 Corners of tiles meet.

5 All tiles are in level.

6 White cement is used for filling the joints

7 All joints are parallel to the adjacent wall.

Check and Allow : Tandur/Shahabad/Marble flooring

Refer drawing : Floor plans

Finishing Schedule

Check the following

1 Material as per approved quality with uniform colour, straight edge and right angles.

2 Thickness of stone is 25mm minimum.

3 Any pattern has been specified.

4 Slopes and levels have been marked.

5 Size of stone is as specified.

6 Centre lines of floor have been marked.

7 The slab has been cleaned thoroughlyof all loose materials and has been wetted.

Following to be Ensured while work is in progress.

1 Bed mix to be 1:2 cement : lime

2 Thickness of bed to be 20mm

3 Stone to be laid on cement slurry of honey dew consistency.

4 Joints to be flushed with cement slurry.

5 All corners meet.

6 All joints are parallel to the adjacent wall.

7 All stones are in level.

8 There is no hollow beneath a stone. Hit with hammer to hear the sound.

Check And Allow : Glazed / Ceramic tiling

Refer drawings : Floor plans

Finishing Schedule

Check the following

1 Pattern specified, if not, follow stretcher born.

2 Tiles are of approved make, grade and color.

3 Thickness of tiles, as specified.

4 Variation in color.

5 Bedding plaster is rough and cured.

6 Levels and slopes

7 The mason has experience in tiling.

8 proper tools for cutting tiles.

9 Corner beads.

10 Edge tiles.

11 All pipes have been embeded.

Following to be Ensured while tiling is in progress.

1 White cement to be used.

2 All vertical joints to be in plumb.

3 Horizontal joints to be in level.

4 Joints of flooring and dado must match.

5 All joints to be parallel to adjacent to wall.

6 Ensure plumbing points are not covered with tiles.

Check and Allow : Brick Bat Coba

Refer drawings : Terrace Plans

Check the following

1 Roof slab cleaned of all loose material with a wire brush.

2 Roof slab roughened and washed.

3 Level marks and slopes identified, by lines on parapet wall. All slopes to be

towards rain water pipes.

———◉———

FOLLOWING TO BE ENSURED while work is in progress.

1 Mix to be 2:2:7 Lime: Surkhi : Over burnt brick bat

2 Leveling to be done with wooden floats.

Check And Allow : Terrace IPS / Tiling

Refer drawing : Terrace plans

1 Check for leakages from slab. Keep terrace flooded with water for 48 hours.

(If leakage observed, repair the same)

2 Surface thoroughly cleaned, roughened and watered.

3 Levels and slopes.

Following to be Ensured while work is in progress.

1 Make slurry of one bag of cement in 9 liters of water and waterproofing

compound.

2 Slurry to be 3mm thick.

3 After slurry sets bedding of 20mm thickness to be laid.

4 Curing to be Flooding.

Check And Allow : Internal Paint / White wash - First coat

Refer Drawing : Floor plan

Finishing schedule

Check the following

1 Plaster has been completed and cured for seven days

2 Flooring skirting / dado completed.

3 Door and Window shutters fixed.

4 Door and Window frames and shutter protected.

5 Electrical wire complete.

6 Electrical switches fixed and protected.

7 Plumbing and sanitary completed.

Check And Allow : Internal Painting / white wash - Final coat

Refer drawing : Nil

Check the following

1 First and second coats completed over a premier coat.

2 All glazing fixed and protected.

3 All flooring polished and protected.

4 All electrical fittings and sanitary wares protected.

5 All locks fixed.

6 External painting complete.

7 Leakage , if any checked and repaired.

Check And Allow : External Painting

Refer Drawing : Finishing schedule
Elevations
Check the following
1 All down water pipes fixed.
2 Plinth protection complete.
3 Street light pipe conduits fixed.
4 Window frames fixed and protected.
5 Terrace waterproofing complete.
6 Pipe joints checked for leaks and tested.
7 O.H. water tanks complete.
8 Terrace and all flats cleaned of all debris.
9 All plaster has been repaired.

<u>Plumbing work</u>

1. check for leakages in G.I. concealed lines before the plastering work.
2. Fix all concealed G.I. lines in position firmly with 40mm plumbing nails
3. Open G.I pipelines painted.
4. Top level of nahani trap before water-proofing work.
5. Keep the nahani trap/ P-trap plugged
6. Keep the opening in G.I pipe lines plugged before starting the internal plaster
7. All ghabdi work and finishing of concealed lines cured properly
8. The heights of the fittings from finished floor level
9. The Water proofing work in bath and W.C. is not damaged during plumbing work.
10. Proper pressure to all the taps and cocks
11. The boiler connections on loft properly plugged
12. Fixing of G.I./C.I./A.C. lines in plumb and clamped properly
13. Leakage in the main inlet and outlet G.I. pipelines of the water tanks
14. P.V.C. outlets pipes for wash hand basin and kitchen sink fixed properly
15. Escape spouts from W.C. and bath are provided.
16. In Hot and cold mixer unit, hot connection should be kept on the left and cold on right.

<u>Drainage work</u>

1 After one coat of water proofing , nahani trap should be fixed with a minimum of 15cm from the side walls.

2 P-trap for W.C. panfix after one coat of water proofing. Adjust the height of commode or W.C. pan with respect to the Finished floor level.

3 Proper slopes to all waste soil pipes and sewer pipes.

4 Outlets of bath, W.C. and kitchen should be in one horizontal plane.

5 All S.W.G pipes should be laid having collar end in the direction of flow and the spigot opposite to the direction of flow.

6 Fix the grating for rainwater pipe on terrace and spout in the balcony.

7 Fix a cowl on ventilating pipe.

8 Gully trap is used between sanitary fittings and the building drain.

9 Gully trap will disconnect a building from a building drain, by means of a 75mm water seal depth. It prevent sewage gases from entering the building.

10 Stone ware intercepting sewer trap is provided at the extreme end of the building drain adjoining the boundary of the premises near the public sewer.

Aluminium windows fixing work

1 Plastered opening at right angles and the diagonals of the opening must be equal.

2 No gap between the window and the plaster. Gap can create leakages.

3 Aluminium sections as per the measurements.

4 Horizontal and vertical members are exactly the right angles.

5 Provide rubber packing on all the sides of the glass panel.

6 clean tracks

7 check the sliding operation of the shutters for free movement.

8 Thickness of glass as specified

9 Quality and operation of rollers at the bottom of the fully glazed shutters.

10 Gaps formed between the walls and windows grouted with epoxy based, rubberized compound. For eg. toughseal or polysulphide using a grouting gun.

11 Handles and locking systems as specified.

12 M.S. aluminium grill from outside for safety purpose.

<u>Electrical work</u>

1 Fan points should be diagonally in the centre of the room.

2 switch boards should be nearest to the entrance.

3 Switches for bathroom and toilet should be outside the door.

4 Power point in the kitchen should be nearest to the kitchen platform.

5 Height of all the switch boards and electrical points as specified.

6 Quality of all the materials is as approved.

7 check the colour codes and sizes of the wire used for all electrical points.

8 Earthing plate electrodes are made up of copper or galvanized iron.

9 Smooth working of all the switches.

10 Check the supply for all the points by megger or test lamp.

11 Check the main supply for sufficient voltage.

12 check all the points in the staircase.

13 All common supply like parking, street lighting are as specified.

14 check the rating of the fuse in each circuit.

15 check the effectiveness of earth continuity using Ohm-meter and a battery.

16 Provide lightning-rod to protect building from lightning.

Elevator works

1. The lift pit should be taken to a hard strata of ground.Depth should be 1.40m below the lowest landing level. Extra depth should be filled with plum concrete.
2. The lift pit should be completely watertight.
3. Centre opening doors are recommened.
4. Provide scaffolding for lift erection.The horizontal supports of scaffolding should be) 0.9m to 1.05m
5. Provide pockets and grout them for rails, brackets, indicator, boxes etc. in position.
6. All R.C.C. beams in lift well should be marked with red oil paint.15cm stripes on top and bottom of beam. This heps in locating the R.C.C. beam position. The rag bolt fasteners can be fixed accordingly in the R.C.C. portion.
7. Check bottom slab of the machine room with all openings is as per the requirements of manufacturer.
8. Hook in the top slab required to set the pulley at the time of erection.
9. Lift machine room flooring should be of I.P.S.
10. Check the rigidity of the foundation concrete and the bolts provided to the lift machine
11. Good quality trap door is fixed to close the opening for the trap door.
12. Check quality of the foundation concrete for buffer springs.
13. The finishing of lift car from inside.

1. Separate electric meter for the lift.
2. Sufficient ventilation in the lift machine room.

3. Weld mesh to act as a guard from birds etc.
4. Functioning of earthing system.
5. Key for the emergency operation of lift.
6. Instruction plate is fixed.
7. Ensure that there is no water seepage in the lift pit, machine room and hoist way.
8. Maintenance of elevators by authorized license agency.

Firefighting work

1. If the height of the building exceeds 15m, building requires

self- arrangements for fire control system.

2 OVER HEAD WATER tank should have an additional capacity of water as per the height and area of the building.

3 Civil works like excavation, making holes in brick walls, concrete and their refinishing is done properly.

4 ISI marked quality of material supplied by the contractor.

5 Check all fire extinguishers are in a working condition.

6 Fire hydrant should be fixed near the accessible road.

Bituminous Road Work

1 APPROVED LAYOUT FOR roads

2 Complete L section of all internal roads for excavation and filling

3 Surface drain off details

4 Slopes and radius at various points

5 Check formation level, excavation level and width of excavation.

6 Ensure that the surface is in proper line, level, slope, camber, curves .

7 Murum layingin two layers. Each layer should be watered and compacted with roller of 10 T capacity.

8 23cm height soling. Roll with 10-12T roller.

9 Metalling 10cm thick. Compacted with 10-12T roller.

10 Check gradient/ slopes.

11 Grouting by bitumen. Semi grouting at the rate of 2.7Kg/sqm.to penetrate a depth of 2.5cm below the surface of the metal. For full grouting at the rate of 10kg/sqm. To penetrate to a depth of 7.5cm below the surface of the metal.

12 Spread the stone grit over the grouted surface. Ensure proper rolling by a road roller.

13 clean the grouted surface with wire bush. Apply the tack coat of bitumen at the rate of 1.5Kg/smt.

14Check the bituminous macadam for uniform mixing.

15 Apply seal coat 12mm thick with 95kg of metal to 5kg of tar. Rolled with 12T roller.

List of Registers to be maintained by the contractor at the site

1. Cement issue register

2. Steel Reinforcement

3. Structural steel

4. White cement

5. Bitumen register

6. All types of flooring materials

7. All types of walling materials

8. All sanitary fittings, piping etc.

9. All doors and windows , iron fittings etc.
10. All types of paints
11. Bulk materials viz. sand, aggregate, rubble, bricks, blocks, murum etc.
12. Waterproofing compounds
13. Timber
14. Cube tests
15. Equipment register
16. Site order book
17. Hinderances to work
18. Secured advance
19. Manpower/Labour register
20. Daily progress Report books
21. Daily material consumption report book
22. Machinery log book
23. Quality test reports
24. Stock register
25. Work order files
26. Bills Inward register
27. Bills Outward register
28. Cheque receipt register
29. Cheque issue register
30. T.D.S. certificate issue register

Lists of standard values/Measurements/Dimensions

1. Trial pit size for Excavation 1.2m X1.2mX depth (depends on hard strata)
2. Safe bearing capacity for Hard Rock – 90 T/sqm.
3. Volumetric mix proportion used for P.C.C. is 1:3:6 (6 inches thick)
4. Curing for P.C.C. 14 days
5. Rubble soling is laid 230mm thick
6. D.P.C. (Damp proof course) is over the plinth walls 6 inches

thick using Mix 1:1 ½:3

7.   1 bag cement = 1.25cft=35litre
8.   Cube size for concrete cube test 150mmX150mmX150mm
9.   Water cement ratio =0.55(for practical workability)
10. De shuttering of columns after 36 hours
11. Curing of columns for 15 days
12. Hacking of columns 50 nos. per sft.
13. De-shuttering of outer beams after 24 hours from the date of slab casting.
14. De-shuttering of internal beams sides after 48 hours.
15. Slab curing for 28 days
16. De-shuttering of slab after 15 days
17. De-shuttering beam bottom up to 3m is 14 days and above 3m is 21 days
18. De-shuttering of slab of span above 6m is 21 days
19. Curing for Brick/stone masonry work is 10 days
20. Mix proportion used to plaster (Cement:sand)

Internal ceiling – 1:4
Internal walls - 1:6
External walls (first coat) - 1:5
External walls (second coat) – 1:5

1.   Curing plaster work for 15 days

1.   Basement waterproofing box type – rough Shahabad stone 25mmto40mm of size 2ft x 3ft are used.

1.   Keep Underground water tank at 60cm above the finished formation level of surrounding land.

1.   Capacity of Underground water tank should be 1.5 times of overhead water tank.

1. A free board of 15cm should be provided for U.G. tank.

1. Use concrete mix grade M20 for U.G.water tank

1. Capacity of O.H.water tank is calculated considering 135litre per person.

1. Minimum free board 150mm for O.H.tank

1. Height from finished floor level for following

Wash basin 35 inches
Cistern (high level) 50 inches

Cistern (low level) 12 inches
Urinal 25 inches
Shower 80 inches
Bucket cock 30 inches
Mixer 30 inches

1. Dimension of bathtub are 1.8m X 0.75m X 0.45m
2. Size of Inspection chamber for two pipes 45cm x 60cm and for more pipes 45cm x 90cm
3. Manholes top diameter minimum 60cm.

Bottom diameter = depth of sewer line
minimum depth 1.5m

1. Heights of fittings for Doors

Eye piece 58 inches from F.F.L.
Handle 48 inches from F.F.L
Night latch 40 inches from F.F.L.
Tadi patti 34 inches from F.F.L.

Unit number plate 70 inches from F.F.L.
Name plate 67 inches from F.F.L.
Aldorp 36 inches from F.F.L.

1.   Electrical points

Living room 2 light points, 1 fan point, 1 plug(5amp),
1 plug point (5amp), 1 bell point, 1 T.V. point,
1 telephone point
Bed room 2 light points, 1 fan point, 1 plug (5amp), 1 T.V.
Point, 1 telephone point
Kitchen 1 light point, 1 fan point, 1 power point(15amp)
1 plug (5amp)
Balcony 1 light point,
Water closet 1 light point
Bath room 1 light point, 1 power point (15amp)
Passage 1 light point

1.   Fuse wire for

Geyser 15amp
Mixer/fridge 5amp
Room(5 to 7 points) 5amp
Plug 2 amp

1.   All doors and windows rolling shutters, collapsible gates etc.
     are measured in sqm. Measurements will be of size of
     openings. The given multiplying factors would be applied to
     arrive at the payable quantity.

Type Multiplying factor
Panelled door 2.5 times
Flush door 2.25 times

M.S. window 1.5 times
Grill (simple) 0.5 times
Grill (design) 1.0 times
Rolling shutter 3.0 times
Corrugated sheets 3.0 times
Collapsible door 3.0 times

SECTION – II

1. <u>CONCRETE QUANTITY FOR FOOTING</u>

<u>FORMULAE:</u>

$$V = L \times B \times D + h/3 \left[A_1 + A_2 + (\text{square root of } A_1 A_2) \right]$$

V = Volume of footing

L = Length

B = Breadth

D = Depth/thickness of rectangular base

H = height of trapezoidal portion

A_1 = Area of bottom portion of footing (L X B)

A_2 = Area of top portion of footing (l X b)

2 <u>Concrete quantity of column</u> = L x B X D

L = length of column

B = Breadth of column

D = Floor to floor height of column

3 <u>Concrete quantity of Beams = L X B X D</u>

L = clear span of beam between two columns

B = breadth

D = depth of beam should be considered up to the top of
the slab.

(Note: For lintel consider bearing on both the ends minimum 230mm)

4 <u>Concrete quantity of slab = L X B X slab thickness</u>

L = length of slab = clear span

B = breadth of slab = clear span

5 <u>Target mean strength of concrete</u>

Formulae

Fm = Fck + (t - t / square root of n) X S

Fck = characteristic strength of concrete

n = number of cube test results

s = standard deviation

t = 1.65

6 Determination of concrete Density

Concrete Density = Average weight of cube in kg

VOLUME OF THE CUBE in Cum.

(Note: volume of standard concrete cube is $3.375 \times 10^{3)}$)

7 Determination of Cement content in concrete mix

cement content (kg/cum) = Density of concrete (kg/cum)

$1 + A/C + W/C$

A/C = Aggregate / cement ratio

W/C = Water / cement ratio

8. <u>Water content in concrete mix</u>

Water content = W/C x cement content

(litre or Kg / cum)

9. <u>Total Aggregate content in concrete mix</u>

Aggregate content = A/C x cement content

(kg/cum) = Y Kg/cum

10. <u>Individual Aggregate content in concrete mix</u>

CAI content = "Y" X % of CAI (20mm)

100

CAII content = "Y" X % of CAII (40mm)

100

FAI content = "Y" X % of FAI (fine aggregate)

100

11. <u>Weigh Batching for conventional concreting</u>

Cement = 50 kg

Water = 50 x 0.55 (W/C) = 27.5 lit.

Fine Aggregate (FAI) = 50 X 6.1 (A/C) x 0.515 (% of FAI)

= 157 Kg

Coarse aggregate (CAI) = 50 X 6.1(A/C) x 0.20 (% of CAI)

= 61 Kg

Coarse aggregate (CAII) = 50 x 6.1 (A/C) x 0.285 (% of CAIII)

= 87 Kg

12 Volumetric Batching for conventional concreting

CEMENT = 50 KG

Water = 50 x 0.55 (W/C) = 27.5 litre

Fine aggregate (FA I) =

1. X 6.1(A/C) x 0.515(% of FAI)

—————————————————————————— +   15%

Bulkage

1.77 (DLBD)

Coarse Aggregate (CAI) =
50 X 6.1(A/C) x 0.20 (% of CAI)

———————

1.44 (DLBD)

Coarse aggregate (CAII) =
50 x 6.1 (A/C) x 0.285 (% of CAIII)

———————

1.39 (DLBD)

(Note : DLBD = Dry loose bulk density of all aggregates)

———————

13 <u>Bulking of Sand</u>

I. Take 250 ml. of glass cylinder
II. Fill up with damp sand up to 200 ml. mark

III. Pour some water. The sand settles to it's actual volume which is less than 200 ml.

IV. Let us assume that it is D ml.

V. Calculate the Bulkage using formula

$$\% \text{ Bulkage} = \frac{200 - D}{D} \times 100$$

(Note : This bulking of sand should be known for proper correction to be applied while calculating dry sand requirement)

14 <u>Silt content of sand</u>

- Take a clean 200 ml. glass cylinder.
- Fill it with sand up to certain mark say X.
- Add water above it up to 200 ml. and shake well.
- Allow a period of two hours for settlement.
- Clean sand will settle at the bottom.
- A layer of silt and clay impurities will be seen on the top of sand.
- Now observe the top level. Say it is Y
- Silt content is given by formula

$$\% \text{ Silt content} = \frac{X - Y}{X} \times 100$$

(Note : Total silt content should not be more than 7 % for good quality sand)

15 <u>water absorption test on metal</u>

- Take 2 Kg. of metal in a container.
- Add water to totally submerge the metal
- Let it remain submerged for 24 hours.
- Weigh the wet metal.
- Note down difference in weight.
- Water absorption can be calculated using formula =

% Absorption = Difference in weight

————————————————-—X 100

Original weight

(Note : for metal to be used in concrete work, water absorption should not be more than 5 %)

1. <u>Water absorption test on Bricks and Blocks</u>

- Take 3 nos. of bricks/blocks
- Weigh each sample and note down weight as W_1, W_2, W_3
- Take average

$$\frac{W_1 + W_2 + W_3}{3} = W$$

- Take clean water in a container and submerge the samples for 24 hours.
- Weigh the wet samples again as W'_1, W'_2, W'_3
- Take average

$$\frac{W'_1 + W'_2 + W'_3}{3} = W'$$

- % absorption can be calculated using formula

$$\% \text{ absorption} = \frac{W' - W}{W} \times 100$$

(NOTE :1. % ABSORPTION should not exceed 20% for good quality Bricks. 2. Water absorption of concrete blocks should not exceed 10 %)

---

1. <u>Water absorption test on mosaic tiles</u>

- Take three tiles at random.
- Note down initial weight of all three tiles.

- Submerge tiles in water for 24 hours.
- Weigh the wet mosaic tiles
- Note down the difference in weight.
- % absorption can be calculated using formula

% absorption = Difference in weight
———————————————- x 100
Original weight

(NOTE : % ABSORPTION should not be more than 10 % for good quality mosaic tiles)

18 <u>Steel quantity</u>

- Weight of bar in Kg/meter should be calculated as $(d^2/162)$ where d is the diameter of bar in mm.

- 'L' for column main steel in footing should be considered as minimum 300mm or as specified.

- For bent up bar add 0.42d in straight length of bar. Where d is the clear depth of beam or slab.

- For cantilever beam anchorage length for main steel should be 69 D or length of cantilever whichever is greater. Where D is diameter of bar in mm.

- Number of stirrups = clear span/spacing +1

- Lap length in column should be 45 D. Where D is the diameter of bar in mm.

- Lap length in beam/slab should be 69D where D is the diameter of bar in mm.

19 <u>Slenderness limits for beams</u>

A simply supported or continuous beam shall be so proportioned that the clear distance between the lateral restaints does not exceed

$60\ b$ or $250\ b^2/d$ whichever is less. Where d is the effective depth of the beam and b the width of the compression face midway between the lateral restraints.

For cantilever, the clear distance from the free end of the cantilever to the lateral restraint shall not exceed $25\ b$ or $100\ b^2/d$ whichever is less.

20 <u>Cover to reinforcement</u>

Reinforcement shall have concrete cover and the thickness of cover shall be as follows :

- To each end of bar not less than 25mm, nor less than twice the diameter of such bar.

- For bars in column not less than 40mm, nor less than the diameter of such bar.

- In case of columns of minimum dimension of 200mm or under and bar diameter do not exceed 12mm, a cover of 25mm may be used.

- For bar in beams not less than 25mm, nor less than the diameter of such bar.

- For tensile, compressive, shear or other reinforcement in a slab, not less than 15mm, nor less than the diameter of such bar.

- For any other reinforcement, not less than 15mm nor less than the diameter of such bar.

21 Reinforcement requirement for structural members

1. Beams

Tension reinforcement

Minimum area of tension reinforcement shall not be less than that given by the following :

$$A_s / b\,d = 0.85 / fy$$

where

A_s = minimum area of tension reinforcement

b = breadth of beam or breadth of the web of T-beam

d = effective depth

fy = characteristic strength of reinforcement in N/mm^2

The maximum area of tension reinforcement shall not exceed 0.04 b D.

COMPRESSION REINFORCEMENT for beam

The maximum area of compression reinforcement shall not exceed 0.04 b D. Compression reinforcement in beam shall be enclosed by stirrups for effective lateral restraint.

<u>Side face reinforcement in beam</u>

Where the depth of the web in a beam exceeds 750mm, side face reinforcement shall be provided along the two faces.

The total area of such reinforcement shall be not less than 0.1 % of the web area and shall be distributed equally on two faces at a spacing not exceeding 300mm or web thickness whichever is less.

<u>Maximum spacing of shear reinforcement</u>

The maximum spacing of shear reinforcement shall not exceed 0.75 d for vertical stirrups and 'd' for inclined stirrups at 45^0. where d is the effective depth of the beam. In no case shall the spacing exceed 450mm.

<u>Minimum shear reinforcement for beams</u>

Minimum shear reinforcement in the form of stirrups shall be provided such that :

A_{sv} / b s_v > or equal to 0.4/ fy

where,

A_{sv} = Total cross sectional area of stirrup leg effective in shear.

S_v = Stirrup spacing

b = breadth of beam or breadth of the web of flanged beam

fy = characteristic strength of the stirrup reinforcement in N/mm^2 which shall not be taken greater than 415 N/mm^2

---

2. <u>Reinforcement in Slab</u>

The reinforcement in either direction in slabs shall not be less than 0.15 % of the total cross sectional area.

The diameter of reinforcing bars shall not exceed 1/8th of the total thickness of the slab

3. <u>Reinforcement in columns</u>

• The minimum number of longitudinal bars in column shall be four in rectangular columns and six in circular columns.

• The bars shall not be less than 12mm in diameter.

• Spacing of longitudinal bars measured along the periphery of the column shall not exceed 300mm

• The cross sectional area of longitudinal reinforcement shall ne not less than 0.8% of the cross sectional area of the column

• Transverse reinforcements in the form of circular rings or polygonal links/lateral ties with internal angles not exceeding 135^0 and properly anchored.

22 Calculation of Labour strength

Labour strength required for various activities during specified period is calculated as illustrated here.

◇ Assume total quantity of plaster is 40,000 sft.

◇ Total days given for plastering is = 20 days

◇ Assume task work of one mason is = 150 sft per day

◇ Therefore total masons required

= 40000 / 150 = 267 masons

◇ therefore masons per day

= 267 / 20 = 13.35 say 14 masons per day

Accordingly, labour strength required per day for other various activities can be calculated.

(Note : Appendix- VIII Table of task work per day of skilled and unskilled labour)

23. Appendix – I weight of building materials

SR. No.	Material	Weight in Kg/ m^3
1	cement	1440 kg / m^3
2	steel	7850 kg / m^3
3	Dry – loose sand	1600 kg / m^3
4	Dry – packed sand	1850 kg / m^3
5	Basalt stone	2900 kg / m^3
6	Dry Earth	1200 kg / m^3
7	Packed earth	1500 kg / m^3
8	water	1000 kg / m^3
9	coaltar	1200 kg / m^3
10	Bricks	1600 to 1900 kg / m^3
11	Glass	2560 kg / m^3
12	Limestone	2650 kg / m^3
13	timber	720 kg / m^3
14	P.C.C.	2240 kg / m^3
15	R.C.C. with 2% steel	2450 kg / m^3
16	Brick masonry	1950 kg / m^3
17	Clay or soil (damp)	1760 kg / m^3
18	Concrete block	1800 kg / m^3
19	Cement mortar	2080 kg / m^3

20 Lime mortar $1760 \, kg/m^3$

21 Marble mosaic tile 25cm x 25cm x 22mm thick 3.2 kg/no.

22 Marble mosaic tile 30cm x 30cm x 25 mm thick 5.2 kg/no

23 Glazed tile 15cm x 15cm x 5mm thick   0.25 Kg/no.

24 Marble stone 3/4 inches thick $2620 \, kg/m^3$

25 Granite stone 3/4 inches thick $2800 \, kg/m^3$

26 Cuddapa 5/4 inches thick $2720 \, kg/m^3$

27 A.C. corrugated sheet $16 \, kg/m^2$

28 Bitumen $1040 \, kg/m^3$

29 Window frame   2.1 kg /no

30 Door frame 3 ft. X 7 ft.   27 kg / no

31 Rubble masonry $2100 \, kg/m^2$

32 G.I. sheet 24 gauge $5 \, kg/m^2$

33 G.I. sheet 16 gauge $10 \, kg/m^2$

24 <u>Appendix – II</u> <u>Geometrical formulae</u>

Sr.no	Geometrical figure	Formulae
1	Rectangle	Area = length x breadth
2	Square	Area = $side^2$
3	Right angled triangle	Area = 1/2 x base x height
4	Triangles	Area = sqare root of {s (s-a) (s-b) (s-c)} where s = a+b+c/2 a,b,c are sides of the triangle
5	Equilateral triangle	Height = (a x 1.73)/2 Area = $(a^2$ x 1.73) /4 a = side of the triangle
6	Circles	Area = 3.14 x r^2 = $(3.14$ x $d^{2)})/4$ r = radius of the circle d = diameter of the circle
7	Circular cylinder	Surface area = 2 x 3.14 x r x (h + r) volume = 3.14 x r^2 x h volume = $(3.14$ $d^2)$ /4 x h h = height of cylinder r = radius of cylinder d = diameter of cylinder
8	Sphere	Area = 4 x 3.14 x r^2 volume = (4 x 3.14 x r^3) / 3 r = radius of sphere

Sr.no	Geometrical Figure	Formulae
9	Regular Hexagon	Area = $(3 \times 1.73 \times a^2) / 2$ a = side
10	Right regular pyramids	Volume = 1/3 x Area of base x height Area = 5/2 x side x (radius of inscribed circle)
11	Right circular cone	Volume = $1/3 \times 3.14 \times r^2 \times h$ r = radius of base h = height
12	Trapezoid	Area = (a + b) / 2 x h a & b = two parallel sides h = height

25 <u>Appendix – III</u> <u>Weight of R.C.C. steel</u> (<u>Reinforcement bars)</u>

Sr. No Diameter of steel Bar Weight per Running meter

Sr. No	Diameter of steel Bar	Weight per Running meter
1	6 mm mild steel	0.22 kg
2	8 mm tor-steel	0.39 kg
3	10 mm tor - steel	0.62 kg
4	12 mm tor - steel	0.89 kg
5	16 mm tor - steel	1.58 kg
6	20 mm tor- steel	2.47 kg
7	25 mm tor-steel	3.85 kg
8	32 mm tor - steel	6.31 kg

26 Appendix - IV Weight of structural steel

Sr. no.	Steel member	Weight/RM
1	ISMB 100 X 75	11.5 KG
2	ISMB 125 X 75	13.0 KG
3	ISMB 150 X 80	14.9 KG
4	ISMB 175 X 90	19.3 KG
5	ISMB 200 X 100	25.4 KG
6	ISMC 75 X 40	6.8 KG
7	ISMC 100 X 50	9.2 KG
8	ISMC 125 X 65	12.7 KG
9	ISMC 150 X 75	16.4 KG
10	ISMC 175 X 75	19.10 KG

27 <u>Appendix – V Weight of M.S. Angle</u>

Sr. No. M.S. Angle size in mm Weight / RM

Sr. No.	M.S. Angle size in mm	Weight / RM
1	ISA 25 X 25X 3	1.1 KG
2	ISA 25 X 25 X 5	1.8 KG
3	ISA 30 X 30 X 4	1.8 KG
4	ISA 30 X 30 X 5	2.2 KG
5	ISA 35 X 35 X 5	2.6 KG
6	ISA 35 X 35 X 6	3.0 KG
7	ISA 40 X 40 X 5	3.0 KG
8	ISA 40 X 40 X 6	3.5 KG
9	ISA 50 X 50 X 5	3.8 KG
10	ISA 50 X 50 X 6	4.5 KG
11	ISA 65 X 65 X 6	5.8 KG
12	ISA 75 X 75 X 6	6.8 KG
13	ISA 75 X 75 X 8	8.9 KG

28 Appendix - VI R.C.C. steel in kg per cubic meter of concrete

SR. No.	Concrete Item	Weight in kg/cubic meter of concrete
1	R.C.C. footing	80 kg/cum
2	R.C.C. Beam	110 kg/cum
3	R.C.C. Column	160 kg/cum
4	R.C.C. Chhajja	50 kg/cum
5	R.C.C. Roof slab	80 kg/cum

(NOTE : ABOVE THUMB rules may be adopted for rough estimation of reinforcement steel)

29 <u>Appendix</u> – <u>VII</u> <u>Consumption</u> <u>Factors</u>

1. Cement

Sr. No.	Items	Unit	Quantity (Bag)
1	P.C.C. 1:4:8	Cum	3.5
2	P.C.C. 1: 3: 6	Cum	4.1
3	R.C.C. 1:2:4	Cum	6.2
4	R.C.C. 1 : 1 1/2 : 3	Cum	8.1
5	R.C.C. 1 : 1: 2	Cum	11.0
6	Masonry U.C.R. 1:6	Cum	1.70
7	Masonry U.C.R. 1:5	Cum	2.05
8	Brick work 1:5	Cum	1.80
9	Brick work 1:6	Cum	1.45
10	Block masonry 200mm thick 1:6	Sq. meter	0.25
11	Block masonry 200mm thick 1:5	Sq. meter	0.30
12	Block masonry 150mm thick 1:5	Sq.meter	0.25
13	Block masonry 100mm thick 1:4	Sq.meter	0.16

14 D.P.C. 40mm thick CM 1:3 Sq.meter 0.40

15 Plaster 20mm thick CM 1:5 Sq. meter 0.12

16 Plaster 20mm thick CM 1:4 Sq.meter 0.14

17 Plaster 20mm thick CM 1:3 Sq.meter 0.16

18 Plaster 20mm thick CM 1:2 Sq. meter 0.25

19 Plaster 25mm thick CM 1:3 Sq.meter 0.22

20 Plaster 25mm thick CM 1:2 Sq.meter 0.33

21 Plaster 15mm thick CM 1:5 Sq. meter 0.09

22 Plaster 15mm thick CM 1:4 Sq.meter 0.11

23 Plaster 15mm thick CM 1:3 Sq.meter 0.12

24 Pointing on Brick work 1:2   Sq.meter 0.09

25 Pointing on Brick work 1:3   Sq. meter 0.06

26 Pointing on Rubble work 1:2 Sq.meter 0.11

27 Pointing on Rubble work 1:3 Sq.meter 0.09

28 Pointing on Rubble work 1:1 Sq.meter 0.14

29 China mosaic flooring   Sq.meter 0.11

30 Shahabad stone flooring   Sq.meter 0.16

31 Tandur stone   Sq. meter 0.16

32 Mosaic tiles   Sq.meter 0.16

33	Glazed tiles	Sq.meter	0.16
34	Indian Patent stone 50mm	Sq.meter	0.22
35	Water proofing by 40mm patent stone	Sq.meter	0.33
36	Water proofing by 25mm to 100mm stone	Sq.meter	0.55
37	R.C.C. Storage tank	100 lit.	0.66
38	Gunniting 40mm to 50mm thick	Sq.meter	0.60
39	Nahani 76cm x 76cm	Per no,	1.0
40	Nahani 92cm x 76 cm	Per no.	1.5
41	Nahani 92 cm x 92cm	Per no.	2.0
42	Jointing SW pipe 150mm	R.M.	0.05
43	Jointing SW pipe 230mm	R.M.	0.09
44	Jointing Hume Pipe 150mm	R.M.	0.07
45	Jointing Hume pipe 230mm	R.M.	0.10
46	Jointing Hume pipe 300mm	R.M.	0.17
47	Jointing Hume pipe 350mm	R.M.	0.20
48	Jointing Hume pipe 450mm	R.M.	0.25
49	Jointing Hume pipe 500mm	R.M.	0.30
50	Jointing Hume pipe 600mm	R.M.	0.33
51	Jointing Hume pipe 700mm	R.M.	0.40
52	Jointing Hume pipe 800mm	R.M.	0.53

53 Jointing Hume pipe 900 mm   R.M. 0.60

54 Jointing Hume pipe 1000 mm   R.M. 0.66

55 Jointing Hume pipe 1100 mm   R.M. 0.82

56 Jointing Hume pipe 1200mm   R.M. 0.89

57 Jointing Hume pipe 1400 mm   R.M. 0.96

58 Brick masonry conical manhole 5 feet deep No.   18

59 For Extra depth   RFt   2.5

60 Inspection chamber 3' x 1'-6" x 1' deep   No.   5.0

61 Circular manhole 8 ft. Dia. 20 ft. deep   No   210.0

62 For extra depth   Rft.   10.0

2. Sand and Aggregate

Sr. No.	Items	Unit	Sand	Aggregate
1	P.C.C. 1:4:8	Cum	0.46	0.92
2	P.C.C. 1:3:6	Cum	0.45	0.90
3	R.C.C. 1:2:4	Cum	0.42	0.85
4	R.C.C. 1 : 1 1/2 : 3	Cum	0.41	0.82
5	R.C.C. 1:1:2	Cum	0.38	0.76
6	U.C.R. CM 1:6	Cum	0.36	
7	U.C.R. CM 1:5	Cum	0.36	
8	Brick work 1:6	Cum	0.31	
9	Brick work 1 :5	Cum	0.36	
10	Block 1:6	Cum	0.27	
11	Block 1:5	Cum	0.26	
12	Plaster 1:5- 20mm	Sq.m	0.02	
13	Plaster 1:4 - 20mm	Sq.m	0.02	
14	Plaster 1:5 - 15mm	Sq.m	0.016	

15   Plaster 1:4-15mm   sq.m 0.016

16   Flooring   sq.m 0.023

3. Miscellaneous

Sr. No.	Items	Unit	Quantity
1	Brick Masonry Bricks – 230mm x 110mm x 65mm	Per cubic meter	460 nos
2	Brick Masonry 230mm Bricks – 230mm x 110mm x 65mm	Per square meter	110 nos
3	Brick Masonry 110mm Bricks – 230mm x 110mm x 65mm	Per square meter	55 nos
4	Block masonry Blocks – 400mm x 20mm x 200mm	Per cubic meter	65 nos
5	Block masonry Blocks – 400mm x 200mm x 150mm	Per cubic meter	85 nos
6	Block Masonry Blocks – 400mm x 20mm x 100mm	Per cubic meter	125 nos
7	Block Masonry - Blocks	Per square meter	13
8	U.C.R. Masonry - Rubble	Per cubic meter	1.3 CUM
9	Mosaic flooring tiles – 250mm x 250mm	Per square meter	16 nos

---

10 Glazed tiles – 200mm x 100mm Per square meter 50 nos

11 Glazed tiles – 150mm x 150mm Per square meter 45 nos

12 Glazed tiles – 100mm x 100mm Per square meter 100 nos

13 Reinforcements – Binding wire Per metric tone   10 Kg

30 APPENDIX – VIII
<u>Concrete Mix – Volumetric Proportions</u>

Sr. No. Concrete Mix Volumetric proportion

Sr. No.	Concrete Mix	Volumetric proportion
1	M5	1 : 5 : 10
2	M7.5	1 : 4 : 8
3	M100	1 : 3 : 6
4	M150	1 : 2 : 4
5	M200	1 : 1 1/2 : 3
6	M250	1 : 1 : 2

(NOTE : CONCRETE MIX proportion of

Cement : Sand : Aggregates for other Mix such as M300 M350, M400,M450,M500 are calculated as per

Mix Design in IS code)

31 Appendix – IX
<u>General Conversion Table</u>

Sr. no.	To Convert	to	Multiply by
1	Gallons	Litres	4.546
2	Pounds of water	litres	0.454
3	Tons per sq. inche	Kg per sq.mm	1.575
4	Horse power	watt	746
5	Cubic CM	Cubic inches	0.061
6	Cubic feet	Cubic meter	0.028
7	Cubic meter	Cubic feet	35.32
8	Cubic meter	Cubic yard	1.308
9	Cubic yard	Cubic meter	0.7645
10	feet	meter	0.3048
11	Gallon	Cubic feet	0.1606
12	inches	Centimetre	2.54
13	inches	millimetres	25.40
14	kilogram	lbs	2.205
15	tons	kilogram	1016.05
16	miles	Kilometres	1.6092

17	Kilometre	metre	1000
18	kilometre	mile	0.62
19	1 metre	centimetre	100
20	1 metre	feet	3.28
21	1 metre	Yard	1.093
22	1 centimetre	millimetre	10
23	1 Yard	feet	3
24	1 mile	feet	5280
25	1 tonnes	kilogram	1000
26	1 kilogram	grams	1000
27	1 quintal	kg	100
28	1 gm	tola	0.086
29	1 sq. metre	sq. cm.	10000
30	1 Acre	Guntha	40
31	1 Guntha	Sq. feet	1089
32	1 Acre	Sq. feet	43560
33	1 hectatre	Sq. metre	10000
34	1 hectre	Acre	2.47
35	1 brass	Sq. feet	100

32 APPENDIX – X <u>Units of Measurement</u>

Sr. No.	Item	Unit for Labour work	Unit for (materials + Labour)
1	Excavation	Cum/Cft	Cum/Cft
2	Back filling/ soling	Cum/Cft	Cum/Cft
3	Footing (concreting)	Number	Cum/Cft
4	P.C.C.	Sq.metre/sq.ft.	Cum/Cft
5	Column/Beam/Lintel Concreting	RM.Rft.	Cum/Cft
6	Slab concreting	Sq.metre/sq.ft.	Cum/Cft
7	Pardi,chhajja , loft concreting	Sq.metre/Sq.ft.	Cum/Cft
8	Staircase concreting	Per steps(No.)	Cum/Cft
9	Brick masonry 15cm thick	Sq.metre/Sq.ft.	Sq.metre/Sq.ft.
10	Brick masonry 230mm thick	Sq.metre/Sq.ft.	Cum/Cft
11	Brick masonry 350mm thick	Cum/Cft	Cum/Cft
12	U.C.R. masonrry	Cum/Cft	Cum/Cft
13	Plastering/pointing	Sq.metre/Sq.ft.	Sq.metre/Sq.ft.
14	Flooring/tiling	Sq.metre/Sq.ft.	Sq.metre/Sq.ft.
15	Skirting/steps	RM/R.ft.	RM/R.ft.

16	Sill Patti	RM/R.ft.	RM/R.ft.
17	Kitchen Otta	RM/R.ft.	RM/R.ft.
18	Painting	Sq.metre/Sq.ft	Sq.metre/Sq.ft
19	Doors / windows	Sq.metre/Sq.ft	Sq.metre/Sq.ft
20	Plumbing pipes	RM/R.ft.	RM/R.ft.
21	Plumbing fittings and Sanitary ware	No	No
22	Water-proofing (terrace)	Sq.metre/Sq.ft	Sq.metre/Sq.ft
23	Water- proofing (bath/w.c.)	No	Sq.metre/Sq.ft
24	Road work (asphalting)	Sq.metre/Sq.ft	Sq.metre/Sq.ft
25	Septic tanks	litre	No
26	Water tanks	litre	litre
27	Electrical points	No	No
28	Electrical wiring	RM/R.ft.	RM/R.ft.
29	Gardening (lawn)	Sq.metre/Sq.ft	Sq.metre/Sq.ft
30	Gardening trees & bushes	No	No

33. Appendix – XI **Labour Task work or work turn out per day**

Sr. no.	Item description	Labour	Quantity/day per labour
1	Excavation in soil	mazdoor	35 cft
2	Excavation in hard soil	mazdoor	75 cft
3	Excavation in ordinary soil	mazdoor	100 cft
4	Brick work in foundation & plinth	mason	55 cft
5	Brick work in super structure	mason	35 cft
6	Partition wall half brick thick	mason	50 sft
7	Random rubble masonry	mason	35 cft
8	Plastering 12mm thick	mason	80 sft
9	White washing	White washer	700 sft
10	Colour washing/painting	painter	2000 sft
11	Painting door/window	painter	250 sft
12	Painting one coat on large surface	painter	350 sft
13	Distempering one coat	painter	350 sft
14	Timber sal/teak framing	carpenter	2.50 cft
15	Timber country wood framing	carpenter	5.0 cft
16	Door/window shutter panelled or glazed	carpenter	7.0 sft
17	Manglore tiling	Tile layer	80 sft

34 APPENDIX – XII
Labour Requirements for different works

SR.no.	Work details	Quantity	Labour	Number
1	Excavation in foundation in ordinary soil up to 100 ft. & lift of 5 ft.	1000 cft	Belders	5
			mazdoor	4
2	Refilling excavated soil in foundation	1000 cft	Belders	3
			mazdoor	2
			bhisti	1/2
3	Disposal of surplus earth	100 cft	Mazdoor	1
4	Laying cement concrete work	100 cft	Belders	2
			mazdoor	3
			bhisti	1/2
			mason	1/2
5	R.C.C. work laying reinforced concrete work	100 cft	Belders	3
			mazdoor	3
			bhisti	1 1/2
			mason	1/2
6	Centering & shuttering flat surfaces	100 sft	Belders	4
			carpenters(class II)	4
7	Reinforcement work	1 quintal steel	Fitter	1
			belder	1

8	Random rubble masonry	100 cft	Mason	3
			belder	3
			mazdoor	2
			bhisti	1/2
9	Partition wall in super structure in CM 1:4	100 cft	Mason	2 1/2
			mazdoor	4 1/2
			bhisti	1/2
10	Frames of doors and windows	6.40 cft	Carpenter	2
			belder	1
11	Shutters panelled or glazed 4 shutters per day	3.0 cft per shutter	Carpenters	15
			belders	4
12	Fixing flat iron hold fast	36 nos per day	Blacksmith	1
			mason	1
			belder	1
13	Fixing M.S. Rods	54 RFT per day	Blacksmith	1
			carpenter class II	2
			mazdoor	3
14	Flooring IPS 4 cm thick	400 sft per day	Mason	5
			belders	4
			mazdoor	3
			bhisti	1

⬥

12	Plastering 12mm thick	400 sft per day	Mason	3
			mazdoor	3
			bhisti	1
13	White washing	600 sft per day	White washer	1
			mazdoor	1

Don't miss out!

Visit the website below and you can sign up to receive emails whenever Priti Dave publishes a new book. There's no charge and no obligation.

https://books2read.com/r/B-A-NIOG-JQEU

BOOKS2READ

Connecting independent readers to independent writers.

About the Author

Author Priti Dave is a Civil Engineer and the author of Six books. She has written books deals with different genre.

Priti Dave is from India. Mumbai is her birth place. Born on 21st October1968. She had a very brilliant academic records. She became civil Engineer in 1990 and worked with various construction organisations in Mumbai. She read many books written by famous Gujarati authors as well as English authors whose writings inspired her to write.